ROYAL
OBSERVATORY
GREENWICH

T0093805

Mars

Patricia Skelton

Royal Observatory Greenwich
Illuminates

First published in 2022 by Royal Museums Greenwich, Park Row, Greenwich, London, SE10 9NF

ISBN: 978-1-906367-94-7

At the heart of the UNESCO World Heritage Site of Maritime Greenwich are the four world-class attractions of Royal Museums Greenwich – the National Maritime Museum, the Royal Observatory, the Queen's House and *Cutty Sark*.

rmg.co.uk

A CIP catalogue record for this book is available from the British Library.

Typesetting by User Design, Illustration and Typesetting
Cover design by Ocky Murray
Diagrams by Dave Saunders
Printed and bound by
CPI Group (UK) Ltd, Croydon, CR0 4YY

About the Author

Patricia Skelton is an Astronomer and Senior Manager of Astronomy Education at Royal Observatory Greenwich. Her passion for astronomy and space exploration can be traced back to two things: Star Trek and seeing the incredible Solar System images captured by robotic spacecraft as a child. Of all the objects in the Solar System, Mars is her favourite – whenever it's up in the night sky, she makes sure to wave hello.

About Royal Observatory Greenwich

The historic Royal Observatory has stood atop Greenwich Hill since 1675 and documents over 800 years of astronomical observation and timekeeping. It is truly the home of space and time, with the world-famous Greenwich Meridian Line, awe-inspiring astronomy and the Peter Harrison Planetarium. The Royal Observatory is the perfect place to explore the Universe with the help of our very own team of astronomers. Find out more about the site, book a planetarium show, or join one of our workshops or courses online at rmg.co.uk.

Contents

Introduction

On a dark and crystal-clear night, with thousands of stars scattered across the sky from horizon to horizon, you cannot help but look up in awe at the beauty of the night sky. Gazing up at the stars is a universal human experience, one that stretches across time. Our innate curiosity about the wonders of the night sky has seen us push the boundaries of science and technology to understand the Universe and our place in it. Of all the questions we have yet to answer, there is one in particular that has occupied our minds for centuries – are we alone in the Universe?

Seeking an answer to this very question has led us to become somewhat obsessed with a cold, rusty and dusty world in our Solar System – the planet Mars.

Of all the astronomical bodies in our Solar System, it is fair to say that none has captured our imaginations as intently as the Red Planet. Known to the ancients as one of the five wandering stars in the night sky, its crimson colour was said to be reminiscent of bloodshed and war. This association led the Ancient Greeks to name the planet Ares after their god of war while, in a similar manner, the Romans named it Mars. The ancient Chinese called it Huoxing and the Japanese called it Kasei, both names meaning 'fire star'.

For millennia the Red Planet remained a point of light in the sky, a destination that we could never reach. The invention of the telescope provided a way to get a closer look at Mars and early telescopic observations revealed the presence of surface features similar to those seen on Earth. As telescope optics improved over time, so too did our views of the planet. Mars became the focus of intense studies with astronomers keen on gleaning as much information as possible. While not all the results of telescopic studies became public, those that did led to Mars firmly embedding itself in the public psyche.

In the 19th and 20th centuries, Mars served as inspiration for some of the greatest works of science fiction. While several stories painted terrifying tales of Martians invading Earth, others told compelling stories of crewed missions to the mysterious Red Planet. While the former may have led a few people to fear Mars, the latter fired up imaginations and caused many to dream of a day when human beings were no longer bound to Earth and could venture out into space to explore the unknown. Dr Robert Hutchings Goddard (1882–1945), the inventor of the liquid-fuel rocket and the man regarded as the father of modern rocket propulsion, was just one of those inspired by these stories. If you're curious about which one led him to become fascinated with space flight, I'll reveal all a bit later.

The dawn of the era of space exploration meant that studies of Mars were no longer confined to telescopes and our exploration of Mars took a giant leap forward. Since the mid-1960s, we have sent numerous uncrewed spacecraft, landers and rovers to Mars in order to learn more about the planet. With each mission we have gained incredible insights into the planet and made new and exciting discoveries. Images returned from these robotic explorers have enthralled us and revealed a world that is different to our own, yet somehow strangely familiar. Considering the number of missions that have been sent to Mars, and the many upcoming missions, you might think that the public would have grown tired of exploration of the planet. It hasn't. You need look no further than the number of followers on the Twitter accounts for NASA's *Curiosity* and *Perseverance* rovers as evidence of how just how much Mars continues to captivate us today.[1]

Considering that the Red Planet is one of the most explored bodies in the Solar System, why do we continue to scrutinise it? Because there is still much left to learn. Every scientific observation we make of Mars helps us to piece together the story of the planet. Over the course of the last century, our understanding has been completely transformed. We now know that Mars was once a warmer planet with standing bodies of liquid water on

---

[1] Yes, the rovers have their own Twitter accounts. If there was ever a reason to join Twitter, this is it.

its surface. A catastrophic climate change turned it into the cold desert-like world we see today. With our home planet as a reference point, liquid water appears to be essential for life to exist and if Mars was once host to liquid water could life have evolved on the planet? The search for evidence of life on the Red Planet, either past or present, is one of the key objectives of current (and future) missions to Mars.

All the information we gather about Mars allows us to prepare for what will be humankind's next greatest adventure – the first crewed mission to another planet in our Solar System. Following a return to the Moon, Mars is the next logical step in our exploration of the cosmos. Astronomer Carl Sagan (1934–96) recorded a message that was included on a special DVD attached to the *Phoenix Mars* lander, which touched down on the surface of Mars in 2008. The message was made for future explorers of Mars and as this extract shows, it sums up why the lure of the Red Planet is hard to ignore:

> *... maybe we're on Mars because we have to be, because there is a deep nomadic impulse built into us by the evolutionary process. We come, after all, from hunter-gatherers and for 99.99% of our tenure on Earth we've been wanderers and the next place to wander to is Mars. But, whatever the reason you're on Mars is, I'm glad you're there. And I wish I was with you.*

In the chapters that follow, we will embark on a journey through time to see how our knowledge of Mars has grown in leaps and bounds over centuries of observing. We'll begin with our ancestors and find out how observations of the motion of Mars in the sky played a vital role in settling the laws that describe motion within the Solar System. Moving forwards in time, we'll learn about the early telescopic studies of Mars and which characteristics could be discerned from a distance. Entering the era of robotic space exploration, we will consider how orbiting spacecraft, landers and rovers have completely transformed our view of the Red Planet. Finally, we will look to the future of Mars exploration.

Ad Martem.

How Do You Solve a Problem Like Mars' Orbit?

If someone were to ask you, 'Was Mars visible in the night sky on 30 October 1938?', you might not know the answer offhand (kudos if you do!), but you would be able to find the answer by using a stargazing app or any other planetarium-style software and setting the relevant date.[2] These apps and programmes are astronomical time machines. Want to

---

[2] Spoiler alert – the answer to the question is yes: Mars was up in the early hours of the morning. And that particular date wasn't chosen at random, as you'll discover in the next chapter.

know where to aim your telescope to view Mars in October 2025 or October 3025? Easy: just enter the relevant date and the software will provide you with the precise coordinates of Mars. Our model of planetary motion, a result of the revolutionary work of early astronomers who were perplexed by the motions of the planets they observed, allows us to calculate the movement of the planets in the future as well as track them in the past. As you'll discover, the Red Planet played a pivotal role in solving planetary motion.

With no light pollution to impede their observing, our ancestors had spectacular views of the night sky. They saw how the stars moved like clockwork in unison across the sky. However, not all stars in the night sky obeyed this rule. Five bright stars were known to traverse the sky through the constellations of the zodiac at their own pace. The Ancient Greeks called them *planētai* or 'wanderers'.

These stars were the naked-eye planets Mercury, Venus, Mars, Jupiter and Saturn. In addition to their meandering nature, they also did something rather puzzling – every now and then, a planet's journey across the sky slowed and then paused. It would then appear to move backwards across the sky for a period of time, after which it paused once more, before resuming its original motion. This puzzling movement was called **retrograde motion**.

At the time, it was believed that Earth was at the centre of the Solar System (and indeed the Universe), with the Sun, Moon and the planets moving in circular orbits around it. This is known as the **geocentric model** of the Solar System. In order to explain the observed retrograde motion of the planets, in the 2nd century CE Alexandrian philosopher Claudius Ptolemy (*c*.85–*c*.165 CE) modified the existing geocentric model. Each planet was said to move around in a smaller circle (called an epicycle), the centre of which moved along a larger circle (called the deferent) around Earth. The combination of these two motions resulted in the apparent retrograde motion of a planet. This revised geocentric model remained unchallenged for over a millennium (see Figure 1).

In 1543, Polish astronomer Nicolaus Copernicus (1473–1543) published a book that revived a model originally proposed

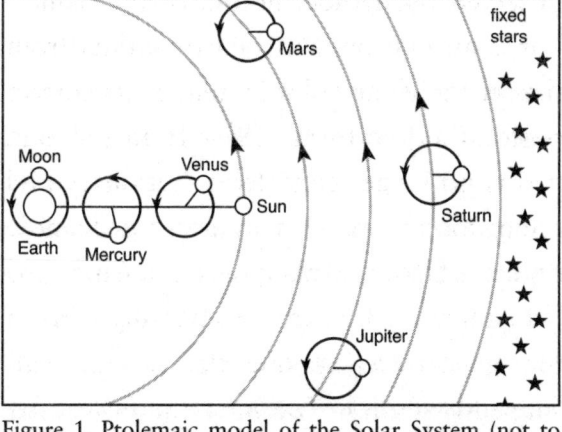

Figure 1. Ptolemaic model of the Solar System (not to scale). The small black circles are the epicycles and the large grey circles correspond to the deferent of each planet. To account for the fact that Mercury and Venus are always near the Sun in the sky, the epicycles of the two planets are centred on a line connecting Earth and the Sun.

well over 1,000 years earlier by the Greek philosopher Aristarchus of Samos (*c.*310–*c.*230 BCE). In this **heliocentric model,** the Sun was placed at the centre of the Solar System with the planets moving in circular orbits around it (see Figure 2). The Copernican model took us one step closer to understanding the complexities of planetary motion and, importantly, it provided a far easier explanation for the observed retrograde motion of the planets – it is an illusion. With Earth shifted from the centre of the Solar System to its correct position as the third planet from the Sun, it was no longer considered stationary and understood to move in an orbit too. So how does this change explain retrograde motion?

Imagine you are competing in a motorsport event taking place on an oval-shaped racetrack. You have the fastest car and you occupy the inside lane. The lane just outside you has been allocated to a slower car. There are staggered starting

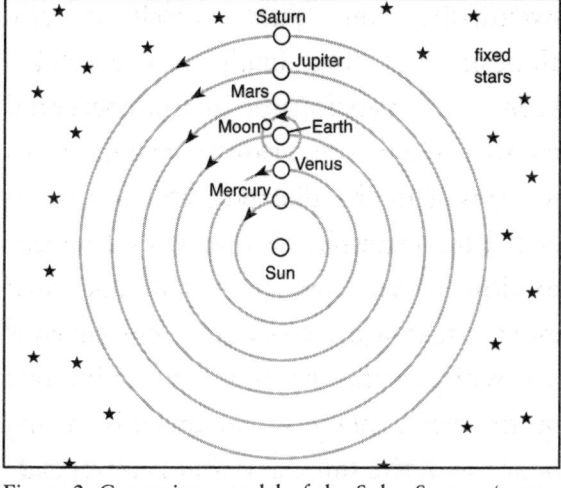

Figure 2. Copernican model of the Solar System (not to scale).

positions on the grid and the slower car is currently further ahead in its lane than you are in yours. Once the lights turn green, you and the driver of the slower car hit your accelerators. You can see the slower car moving and, although you are moving too, it is still ahead of you. It may not feel like it, but you are catching up. As you head into the bend, you can see that you're getting closer and closer to the other car.

14

Eventually, you find yourself directly alongside it before, finally, you overtake. From your perspective, if you measured the movement of the slower car relative to something in the distance, an advertising board for example, the car would appear to slow down, come to a halt and then move backwards. Now just replace your car with Earth, the other car with one of the outer planets and the advertising board with the stars, and that is retrograde motion (see Figure 3).

While Copernicus' heliocentric model was a big step forward, not all astronomers were keen on the idea of placing the Sun at the centre of the Solar System, including Danish astronomer Tycho Brahe (1546–1601). Tycho proposed a hybrid of the geocentric and heliocentric models – in his 'geoheliocentric' model, the Sun and the Moon orbited Earth, while the other planets orbited the Sun. Tycho was a brilliant observational

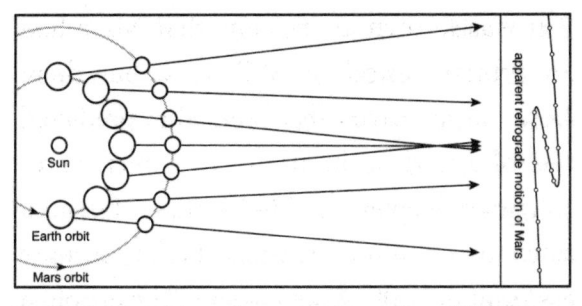

Figure 3. The observed retrograde motion of Mars is an illusion produced because both planets are orbiting the Sun at different speeds. Arrows point to the position of Mars in the night sky as observed from Earth for the different orbital alignments of the two planets (not to scale).

astronomer and studied stellar and planetary positions extensively, collecting data that are considered by many to be the most accurate data obtained before the invention of the telescope. Based on his data, Tycho noticed that there were discrepancies between his observed planetary positions and those predicted by the Ptolemaic and Copernican models of the Solar System. But the motion of one planet was seemingly impossible to understand – that of Mars.

It was known at the time that Mars had an orbital period of 687 days, but there were large errors between the predicted and observed positions of the planet. Enter Johannes Kepler (1571–1630), a German astronomer who became Tycho Brahe's assistant in 1600. Asked to solve the problem of Mars' orbit, Kepler began working through Tycho's observations of the Red Planet (it has been said that Tycho assigned Kepler this task in order to keep him busy, while Tycho himself worked on refining his own Solar System model).[3] What Kepler assumed would take eight days to solve took him a tad longer… eight years to be precise. Throughout this time, Kepler performed numerous painstaking calculations but struggled to make Tycho's observational data relating to Mars match up with a circular orbit in the Copernican model.

---

[3] Initially, Tycho only granted Kepler access to some of his observational data. After Tycho's death in 1601, Kepler gained access to all the data (although some say that he stole it…).

17

The breakthrough came when Kepler found that if the shape of Mars' orbit was changed to that of an ellipse (a squashed circle), he could match the predicted positions of Mars to its observed positions. Kepler had finally solved the problem of Mars' orbit and established the first of his three fundamental Laws of Planetary Motion: the planets move in elliptical orbits around the Sun. By replacing the circular planetary orbits of the Copernican model with elliptical orbits, Kepler revolutionised our understanding of planetary motion. He would go on to use Tycho's observations of Mars as the foundation for his Second Law: a planet moves faster in its orbit when it is closer to the Sun and slower when it is further away. The point at which a planet is closest to the Sun is called **perihelion,** while the point of greatest separation is called **aphelion**.

Eccentricity is a measure of how much an ellipse deviates from a perfect circle and takes a value between 0 and 1. If a planet were to move in a perfect circle around the Sun, it would have an orbital eccentricity of 0. The larger the orbital eccentricity, the more elongated (or elliptical) the planet's orbit. It should be noted that Mars has the second most elliptical orbit of all the planets in the Solar System, with an orbital eccentricity of 0.094.[4] Had Tycho given Kepler data on any other planet, say Jupiter, which has an eccentricity of 0.048, Kepler would likely have taken much longer than eight years to formulate his Laws of Planetary Motion, if at all.

As for Mars itself, it was still nothing more than a red-coloured point of light in the sky when Kepler published his work. But that was all about to change thanks to the invention of the telescope.

---

[4] Mercury takes first place with an orbital eccentricity of 0.206. For comparison, Earth has an orbital eccentricity of 0.017.

Invaded by Mars

Technology has always played a vital role in advancing scientific knowledge and the invention of the telescope in the 17th century is no exception. Italian astronomer Galileo Galilei (1564–1642) is considered by many to be the 'father of modern science' and he was the first person to use a telescope to observe celestial objects. More importantly, he also recorded his observations. In 1610, he was the first to view Mars through a telescope. The limitations of the optics of his telescope meant that he was just about able to see it as a disc but it was, nonetheless, an important observation. It confirmed that Mars was indeed a world of its own in the Solar System and the age of telescopic observations of the Red Planet had begun.

When Dutch astronomer Christiaan Huygens (1629–95) used a telescope of his own design to observe Mars in 1659, he noticed a dark spot on the planet's surface. Huygens proceeded to make a series of sketches of the planet over the course of a few hours and noticed that the location of the dark spot appeared to change. By measuring how long it took for the dark spot to return to its initial position, Huygens estimated that Mars had a rotational period of 24 hours.

The length of a day on Mars was further refined by the Italian astronomer Giovanni Cassini (1625–1712), who observed the planet in 1666. Using the dark Martian surface features that he could see, Cassini determined that Mars rotated slightly slower than Earth and calculated a rotational period of 24 hours 40 minutes. This is remarkably close to the currently accepted value of 24 hours, 39 minutes and 35 seconds.

Although telescope optics gradually improved over the years, observing Mars remained a challenge because it still appeared quite small, which made it difficult to discern individual surface features. To counter this, astronomers tended to concentrate their observations during the period when Mars reached **opposition**. At opposition, Mars and the Sun are on directly opposite sides of Earth which means that Mars lies above the horizon for most of the night (see Figure 4). Oppositions of Mars occur about every 26 months, but if Mars reaches opposition at roughly the same time that it reaches perihelion, then Mars is said to be at perihelic opposition. These extra-close oppositions occur every

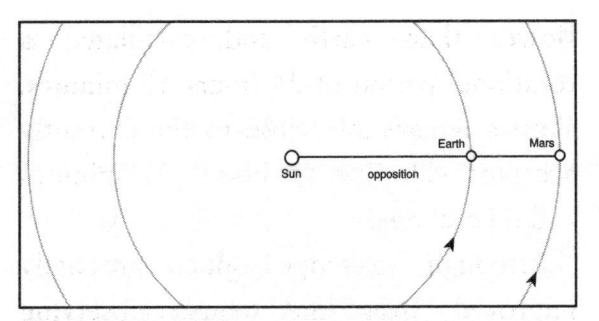

Figure 4. When Mars is at opposition, it is exactly on the opposite side of Earth from the Sun (not to scale).

15 to 17 years and it is at these times that Mars will appear at its biggest and brightest through a telescope.[5]

During the perihelic opposition of 1672, Huygens noticed a white spot located at the south pole of Mars and, in 1704, French-Italian astronomer Giacomo Maraldi (1665–1729), a nephew of Cassini, also observed a white area at the same pole. By studying Mars over the course of the next few oppositions, he spotted

---

[5] The next perihelic opposition of Mars will occur in 2035.

that the size of the white spot varied over time. When Mars next reached perihelic opposition in 1719, Maraldi could no longer see the white spot at all.

British astronomer Sir William Herschel (1738–1822) observed Mars between 1777 and 1783. His 1784 paper, 'On the remarkable appearances at the polar regions on the planet Mars, the inclination of its axis, the position of its poles, and its spheroidal figure; with a few hints relating to its real diameter and atmosphere', presented the results.[6] He determined that Mars had an **axial tilt** of around 28 degrees (not too far from the current value of 25.2 degrees), the planet had an atmosphere and he surmised that the white spots on Mars were polar regions, similar to those we have on Earth (see Figure 5). Due to its axial tilt, Mars would have a

---

[6] On reflection, I should have put a bit more effort into the title of this book...

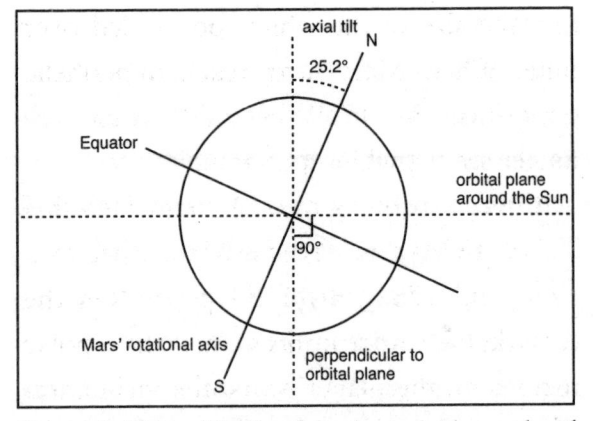

Figure 5. The axial tilt of a planet is the angle of the rotation axis measured relative to a line drawn perpendicular to the orbital plane of the planet. Mars has an axial tilt of 25.2 degrees (for comparison, Earth's axial tilt is 23.4 degrees).

seasonal cycle but the Martian seasons would last longer than the seasons on Earth due to the Red Planet's longer orbital period. Just as Earth's polar regions undergo seasonal changes, Herschel proposed that the Martian polar regions experienced seasonal changes too and that would explain the variations in the size of the white regions that had been observed by Maraldi.

By the late 18th century, astronomers had determined that Mars had semi-permanent features on its surface as well as features that were variable and were attributed to being atmospheric in nature. The next step in the study of Mars was to create a map of the planet showing its surface features and the first people to do just this were the German astronomers Wilhelm Beer (1797–1850) and Johann von Mädler (1794–1874), who published their Martian map in 1840. It included a grid of coordinates and areas of particular interest were marked by letters of the alphabet. The small circular feature Beer and von Mädler labelled 'A' was what they defined as the Martian prime meridian and they used it to help measure the rotational period of the planet. Over the course of the next few decades, astronomers took advantage of the improvement in the quality of telescope optics to produce increasingly detailed Martian maps. They noted the numerous darker regions on Mars and how the shapes and sizes of these regions appeared to vary, with some changes appearing to follow the seasons. Many believed that these areas were seas, while the lighter sections were landmasses; a few even speculated that the darker surface shades were produced by patches of vegetation.

Perhaps the most influential map published in the 19th century, and the one that would spark the obsession about life on Mars, was one produced by Italian astronomer Giovanni Schiaparelli (1835–1910) in 1877. Schiaparelli observed Mars during the perihelic opposition of that year and produced what were then the most detailed drawings of the planet. His map showed darker regions with well-defined boundaries and new surface features that had never been seen before, for which he also introduced a new naming scheme inspired by Earth's geography and classical literature. Perhaps the most surprising details that appeared on Schiaparelli's Martian map were numerous narrow lines that he called *canali*, meaning channels. Unfortunately, *canali* was mistranslated into English as 'canals' and it was this seemingly small error that led to the idea of life on Mars. Why? Because channels occur naturally here on Earth, while canals are manmade. It followed, therefore, that the appearance of canals on Mars could only be the result of a civilisation on the planet. One person, in particular, took this idea and ran with it: the American astronomer Percival Lowell (1855–1916).

In 1894, Lowell set up an observatory with the sole purpose of observing Mars during the perihelic opposition that was to take place that year. Having compiled hundreds of drawings of the planet, Lowell published his work in 1895. Not only did he confirm the existence of the canals, but he also went on to suggest that they were an irrigation system set up by Martians to save their dying planet (see Figure 6). The Martians were diverting water from the polar ice caps to grow vegetation. Needless to say, Lowell's publication fired the imagination of many people, including one Herbert George Wells (1866–1946), an English writer who authored the science fiction novel *The War of the Worlds*. Published in 1898, the story told a terrifying tale of Martians invading Earth and became an instant bestseller.[7] It was this very story that inspired Robert Hutchings Goddard to dream of spaceflight.

While the public became fascinated by the prospect of life on Mars, not all

---

[7] On 30 October 1938, Orson Welles' radio production of the H.G. Wells classic was so realistic that it led many listeners to believe that a terrifying Martian invasion of Earth was underway.

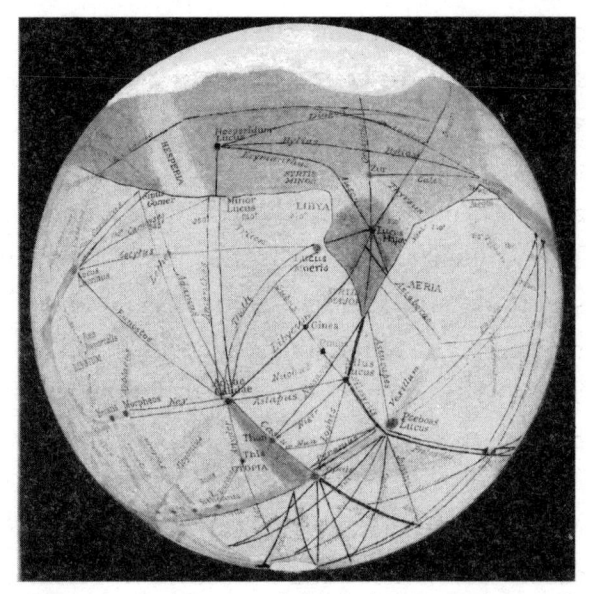

Figure 6. A drawing of Mars by Percival Lowell showing the infamous canals, 1905. *Lowell Observatory Archives*

astronomers were convinced by Lowell's canal theory. British astronomers Edward (1851–1928) and Annie Maunder (1868–1947) argued that the canals were nothing more than an illusion. In 1903, while working at the Royal Observatory in Greenwich, London, they set up a few simple experiments. They covered a series of discs of various sizes with Martian surface patterns that had been augmented with a series of dots. They seated students at different distances from the discs and asked them to draw what they could see. The students closest to the discs could see the finer surface features, including the dots, and rendered them distinctly. Students further away were unable to resolve the individual marks and instead merged them into a series of lines. These results confirmed what the Maunders had suspected – the 'canals' were a trick of the eye. Other astronomers reached the same conclusion, but this did not deter Lowell, and indeed other astronomers, who persisted for years to come with their arguments that there were canals on Mars.

During the first half of the 20th century, astronomers worked hard to discern as much as they could about the physical properties of Mars. In 1924, Americans Edison Pettit (1889–1962) and Seth B. Nicholson (1891–1963) used an instrument called a thermocouple to obtain measurements of the temperature of the planet, yielding values of around 7 °C at the equator and a chilly -68 °C at the south polar ice cap. Further measurements in 1926 implied that the surface temperature plunged well below -75 °C during the Martian night, indicating that the planet's atmosphere was very thin. As for the composition of the atmosphere, astronomers studied the sunlight reflected by Mars to determine what gases were present in its atmosphere (this technique is known as **spectroscopy**). Early measurements suggested an atmosphere similar to Earth's, but, in 1910, results obtained by American astronomer William Campbell (1862–1938) proved conclusively that the Martian atmosphere is markedly deficient in both water vapour and oxygen.

Observations obtained in 1933 supported this and, based on these measurements, German-American astronomer Rupert Wildt (1905–76) suggested that the rest of the planet's oxygen was chemically bound to its surface in the form of iron oxide (rust), which also gave the planet its distinctive red colour. In 1947, Dutch astronomer Gerard Kuiper (1905–73) detected carbon dioxide in the Martian atmosphere for the first time, but the full chemical composition of the atmosphere was still unknown.[8]

Despite the extraordinary progress that had been made by this stage, there was a limit to how much could be learnt about Mars from a distance. But thanks to Robert Goddard's dream, our studies of the Red Planet would soon no longer be bound to telescopes on Earth.

---

[8] The Kuiper Belt, a doughnut-shaped region of icy bodies beyond the orbit of Neptune, is named for Gerard Kuiper.

Is There Life on Mars?

The dawn of the era of space exploration marked a momentous point in human history and the beginning of our journey into the cosmos. Destinations once believed to be beyond our reach could now be explored in unprecedented detail thanks to robotic spacecraft. In the case of Mars, a spacecraft would be able to obtain much-needed scientific data as well as capture close-up views of the surface of the planet. The latter would finally help to settle the debate about the canals, and life, on Mars.

Roughly every 26 months, the orbital positions of Earth and Mars around the Sun are such that a spacecraft can reach the Red Planet with the least fuel possible. The trajectory a spacecraft follows is called the **Hohmann Transfer orbit** and, perhaps counterintuitively, you would never aim towards Mars at the time of launch. Instead, the course of the spacecraft is set to intercept the Red Planet along its orbit. The best way to imagine this is to think of Mars as a distant moving target and the spacecraft as something you want to throw at it. If you aim directly at Mars, your spacecraft will miss it because it will have moved on. Instead, you aim your spacecraft ahead of Mars so that the two meet up. Several course corrections might be required along the way, but, if all goes to plan, a spacecraft will reach Mars around 7 to 9 months after launch (see Figure 7).

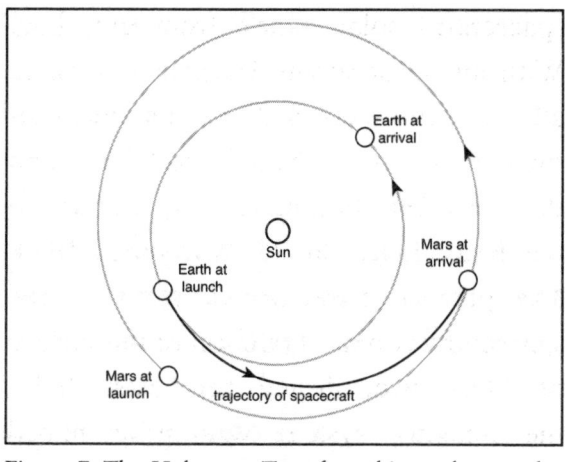

Figure 7. The Hohmann Transfer orbit used to send a spacecraft from Earth to Mars (not to scale).

The early 1960s saw numerous attempts by the Soviet Union to get a spacecraft to Mars, but each ended in failure. In 1964 the United States entered the race to reach the planet with their *Mariner 3* and *4* spacecrafts, both scheduled to launch from Cape Canaveral, Florida, in November that year. *Mariner 3* was the first to launch but, shortly after lift-off, the protective shroud encasing the spacecraft failed to open as intended and prevented the spacecraft's solar panels from unfurling. With an unsuccessful *Mariner 3* mission, all eyes now turned to its identical twin. *Mariner 4* had better luck and the spacecraft began its long journey to the Red Planet on 28 November 1964. The planned trajectory would take the spacecraft within 10,000 km of the surface of Mars and, during this close flyby, the imaging system was programmed to acquire a series of photographs of the Martian surface. *Mariner 4* was equipped with scientific instrumentation designed to measure a number of things, from monitoring the number of **micrometeorite** impacts sustained during its flight through space to determining the composition of the Martian atmosphere.

On the night of 14/15 July 1965, after an eight-month journey through space, *Mariner 4* made history when it became the first spacecraft to perform a successful flyby of Mars and the first to return close-up images of the planet.[9] It captured more than 20 photographs of Mars and began transmitting the images back to Earth, where a team of scientists and researchers were nervously awaiting their arrival. As the data stream from *Mariner 4* trickled in, mission scientists were so anxious to see what the first image would look like that they opted to manually process the image themselves rather than wait for the official computer-processed version. The image data was converted into a series of numbers corresponding to a greyscale (the *Mariner 4* camera took black and white images). These numbers were printed onto strips of paper that were then mounted side by side onto a

---

[9] The first spacecraft to fly by Mars was the Soviet Union's *Mars 1* mission, which achieved this milestone in 1963. Due to a radio failure, however, communication was lost with the spacecraft before it reached the planet.

display panel. Using a set of coloured pastels, the scientists assigned colours to the numbers and hand-coloured the strips to create the first close-up image of Mars. The result of their artistic efforts was later displayed at NASA's Jet Propulsion Laboratory in California (see image 1). The groups involved in the *Mariner 4* mission were both ecstatic and relieved – the image confirmed that the spacecraft was operational and that the camera had locked onto the planet during the flyby. The first image showed the **limb** (edge) of Mars surrounded by the blackness of space. As the rest of the images were received and processed, scientists were eager to see what the surface of Mars looked like. Would there be evidence of canals and civilisation?

The answer was no. The images returned by *Mariner 4* quashed any hopes of Mars being a habitable world with life on it. Instead of a planet covered in vegetation, bodies of liquid water or signs of civilisation, the images revealed a heavily cratered surface similar to the lunar surface, which had, at this point, been imaged many times (see image 2). The impact craters were produced by meteorite and asteroid strikes and, despite the low quality of the images, scientists estimated that the ages of the craters ranged from hundreds of millions to billions of years old. The sharpness of some of the crater walls suggested that these regions of Martian surface had remained pretty much unchanged since they formed. This in turn showed that the craters were untouched by weathering processes such as strong winds, rain and flowing water that would have smoothed and eroded their walls over time. Scientific measurements indicated that the atmosphere on Mars was indeed thin, so thin in fact that the atmospheric pressure at the planet's surface was only around 1% of that at sea level on Earth. To experience Martian surface air pressure here on Earth you would need to ascend to an altitude of 45 km above the ground, something you could not survive unless you wore a pressurised spacesuit.[10] In terms of the atmospheric composition, the results indicated that the atmosphere was predominantly composed of carbon dioxide. The daytime surface temperature was estimated to be a frigid -100 °C and no evidence of a magnetic field around the planet was detected. Mars was not looking very hospitable at all.

Mariner 4 mapped around 1% of the Martian surface and the success of the mission led to subsequent flyby missions being sent to the planet. The twin spacecrafts *Mariner* 6 and 7 flew past Mars in 1969 and returned just over 200 images, with each spacecraft imaging different

---

[10] Please do not try this at home.

regions. Tantalisingly, a few images showed that not all areas on Mars were heavily cratered, some were featureless. As for the proposed canals, images from these missions proved conclusively that neither they, nor any other artificial structures, existed on the planet. These results finally signalled the death knell for the theory sparked by Giovanni Schiaparelli's map almost 100 years previously.

The *Mariner* 4, 6, and 7 missions transformed our view of the Red Planet from being a cold but potentially habitable world to a hostile one devoid of life. Combined, the missions had mapped only around 10% of the surface of Mars and there was a strong drive to send an orbiter to obtain a global view of the planet. *Mariner* 8 and 9 were the final Mars missions of NASA's *Mariner* programme and were designed to be the first Mars orbiters. Each one was fitted with a rocket engine and loaded with fuel to slow down the spacecraft so that it would be captured into orbit around the planet.

Mariner 8 was lost during launch but *Mariner 9* had a successful launch on 30 May 1971 and, on 14 November that same year, it became the first artificial satellite of Mars. Scientists were eager to begin mapping the Red Planet but, as luck would have it, *Mariner 9* entered orbit at the exact time that Mars was enveloped by a global dust storm, concealing the surface from view. The spacecraft (and the mission scientists back on Earth) had to wait until January 1972 for the atmosphere to clear enough for mapping to begin. What was revealed was so striking that it profoundly changed our view of Mars.

Surveying 85% of the planet's surface, *Mariner 9* revealed that it was far more varied than previous flyby images had shown. Massive volcanoes towered above the Martian surface in the planet's northern hemisphere, a system of canyons stretched across the ground for thousands of kilometres along the equator and, perhaps most exciting of all, etched into the surface of Mars were dried riverbeds and valleys, evidence that water had once flowed across the planet (see images 3 and 4). If you're wondering why these features weren't spotted by the previous flyby missions, it was simply because their trajectories took them over other regions of the planet, where craters predominate. As for the darker regions that had been observed on Mars, they weren't patches of vegetation, but regions where the fine red dust covering the surface had been blown away by Martian winds, revealing a darker subsurface. *Mariner 9* also provided us with the first close-up views of the two moons of Mars, Phobos and Deimos.[11] Resembling misshapen potatoes,

---

[11] Phobos and Deimos were discovered by American Astronomer Asaph Hall (1829–1907) in 1877. The names come from Greek mythology.

their surfaces also littered with impact craters, both moons are much smaller than our Moon; Phobos has a diameter of just over 22 km, while Deimos measures just over 12 km across.

What was clear from *Mariner 9* was that Mars was a far more complex world than scientists could have imagined. The startling evidence that Mars once had water flowing on its surface reignited the debate about the existence of life on Mars. If Mars had once been habitable, could there have been life on it? If there was, was it possible that the life evolved and adapted to survive the processes that transformed the Red Planet from a wet and possibly warmer planet to the cold and dry world we see today? To answer this question, we would need to study the surface of Mars from the ground itself.

Once again, the Soviet Union led the way and was the first to land a spacecraft on Mars when its *Mars 3* lander touched down on the planet on 2 December 1971. Despite the safe landing, the lander only returned 20 seconds of data before transmission stopped abruptly, although its partner orbiter continued to obtain data until 22 August 1972. NASA's *Viking* project marked the first attempt by the United States to place landers on the surface of Mars. Similar to the *Mariner* missions, the *Viking* mission consisted of two identical spacecrafts *Viking 1* and *Viking 2*. Each spacecraft was made up of an orbiter and a lander. On 19 June 1976 *Viking 1* entered orbit around Mars and initially spent a month imaging the Martian surface to identify potential landing sites for both landers. On 20 July 1976 the *Viking 1* lander separated from the orbiter and successfully touched down on the surface of Mars at Chryse Planitia, a smooth plain in the planet's northern hemisphere (see Figure 8). Just minutes after touchdown the lander returned

Figure 8. An illustration of the *Viking* landing sequence. The use of a supersonic parachute and retro-rockets slowed the lander to a speed of around 7 km per hour, allowing it to achieve a soft landing. *NASA*

humankind's first glimpse of the surface of Mars. It was a landmark moment in our exploration of the Red Planet – we were no longer restricted to orbital or telescopic views, we now had eyes on the ground.

Viking 2 successfully entered Mars orbit on 7 August 1976 and its lander touched down at Utopia Planitia, a large plain located in an impact basin in the northern hemisphere, in September that year. Both landers revealed a rust-coloured, dry and dusty world lying beneath a reddish-pink coloured sky. The *Viking 2* landing site was scattered with rocks, the majority of which appeared to be pitted with small holes, or **vesicles**. The rocks are thought to be volcanic in nature because of their resemblance to substances on Earth that have been produced by volcanic activity. In stark contrast to this rock-filled region, the site of *Viking 1*'s landing consisted of windswept sand dunes interspersed with rocks (see image 5). The landers were equipped with a range of instrumentation, including a seismometer to record vibrations in the ground, a weather station and a biological laboratory. The purpose of the laboratory was to look for signs of life in the Martian soil by searching for organic matter or evidence of photosynthesis. One of the experiments returned a positive result, prompting excitement among many researchers, but others failed to detect any organic compounds and the outcome of the search for life was ultimately said to be inconclusive. As for the chemical composition of the soil, samples obtained from both landing sites revealed it to be rich in iron, with iron oxide giving the planet its vibrant red colour.

Surface temperatures at the landing sites varied from as low as -120 °C at night to around -20 °C during the day. Other measurements obtained by the instrumentation on board the landers showed that the surface of Mars is bathed in harmful ultraviolet radiation from the Sun due to the thinness of the Martian atmosphere. This radiation effectively sterilises the surface of the planet, making it highly unlikely that life could exist on the ground.

While the landers operated on the surface, the *Viking 1* and *2* orbiters captured over 52,000 images of Mars and provided breathtaking views of canyons, dust storms and volcanoes. The orbiter images were used to produce the highest resolution map of Mars available at the time. It revealed that the planet's southern and northern hemispheres have completely different topographies, with the former dominated by rocky, cratered highlands, while the latter is smoother and lies lower in elevation.

The plan was for the *Viking* landers to perform operations for 90 days after landing but both continued well beyond this. The last transmission from the *Viking 1* lander was made in 1982 and the last data from the second lander was received in 1980. Running low on propellant used to maintain the spacecraft's orientation, the *Viking 1* orbiter was shut down in 1980 having completed its 1,489th orbit of Mars. *Viking 2* developed a leak in its propulsion system and reached the end of its mission in 1978 having orbited Mars 706 times. The fate of the *Viking* orbiters is unknown – it's possible they're still in orbit or have already crashed into the Red Planet. The *Mariner* and *Viking* missions to Mars made enormous impacts on our studies of the planet and inevitably raised more questions. To answer these would require further missions to Mars, but there would be an almost 20-year gap before we returned to the Red Planet.

The Next Generation

The late 1980s and early 1990s saw the launch of three missions to Mars. Unfortunately, they were all unsuccessful – the *Phobos 1* orbiter (Soviet Union) was lost en route to Mars and the *Phobos 2* probe (Soviet Union) was lost near its namesake. *Phobos 2* had managed to gather some data on the Martian moon and was making a close approach to release two landers when communication with the spacecraft ceased. The *Mars Observer* (United States) spacecraft suffered a similar fate when scientists were unable to contact it prior to it being placed into orbit. Mars, it would seem, was reluctant to give up its secrets.

In 1996, NASA tried its luck again and launched two new missions. The first to begin its journey was *Mars Global Surveyor* (*MGS*), an orbiter purpose-built to map the topography of Mars, study weather phenomena and determine the planet's internal composition. *Mars Pathfinder*, which set off from Cape Canaveral just under a month later, was an ambitious mission and consisted of a lander and a robotic rover, called *Sojourner*.[12] *Pathfinder* was designed as a **technology demonstration** because the mission was not only going to test the viability of using rovers to study Mars, but also a new method of delivering a lander to the surface of the planet.

---

[12] Spoiler alert and fun fact: *Pathfinder* and *Sojourner* play an important role in the novel and 2015 film *The Martian*.

In the case of the *Viking* missions, landers were released to the surface from orbiters, but *Pathfinder* would enter the Martian atmosphere directly. Using a parachute to slow its descent, once the lander was 300 m above the surface of the planet, a multiple-lobed airbag system would be deployed, allowing the lander to bounce along the surface before coming to a rest (see Figure 9). This new delivery method meant that an area that was considerably rougher than those chosen for the *Viking* missions could be targeted

Figure 9. An artist's impression of *Pathfinder* shortly after it came to rest on the Martian surface. *NASA/JPL-Caltech*

by *Pathfinder*. Although *Pathfinder* launched after *MGS*, its trajectory allowed it to reach Mars ahead of the orbiter. On 4 July 1997, *Pathfinder*, with *Sojourner* stowed safely on board, began its nail-biting descent to the Martian surface. The chosen landing site was an ancient and rocky flood plain called Ares Vallis, the ultimate test of the new landing system. Mission scientists waited for confirmation that *Pathfinder* and its payload had survived the bumpy landing, and anxiety was soon replaced by celebration when a radio signal confirmed a successful touchdown on Mars. Following the landing, the *Pathfinder* lander was formally renamed the Carl Sagan Memorial Station. Carl Sagan (1934–96) was an astronomer and a passionate science communicator with a skill for making astronomy accessible to the public. He was a consultant and adviser to NASA in the 1950s and played leading roles in a number of programmes including *Mariner* and *Viking*. As a tireless promoter of the exploration of Mars, the renaming of *Pathfinder* was a fitting tribute to him.

After the lander unfurled its petal-shaped solar panels, *Sojourner* trundled onto the Martian surface and, in doing so, became the first rover to explore Mars. Due to the communication delay between Mars and Earth, spacecraft and rovers cannot be controlled in real time,[13] so *Sojourner* operated in a semi-autonomous manner. About the size of a microwave oven, the solar-powered rover was supposed to embark on a seven-day mission but operated well beyond this

---

[13] Although communication signals travel through space at the speed of light, the vast distances involved and the varying distance between Earth and Mars (remember both planets are moving along their respective orbits at different speeds) mean that a signal can take anywhere between 3 and 23 minutes to travel from Earth to Mars and vice versa.

period and spent 83 days exploring the area around its landing site (see image 6). Images returned by both *Pathfinder* and *Sojourner* revealed the presence of rounded rocks, pebbles and cobbles, providing yet more evidence that water once flowed on the surface. Clouds made of water-ice crystals were spotted in the early morning sky and **dust devils,** towering columns of swirling air and dust, were seen gliding across the ground. *Pathfinder* discovered that the airborne dust on Mars is magnetic and most likely contains the mineral maghemite (magnetic iron oxide).

With *Pathfinder* and *Sojourner* safely on the ground, next up was *MGS*, which successfully entered Mars orbit in September 1997. Equipped with high-resolution cameras that could spot objects a few metres across on the surface, *MGS* sent back awe-inspiring images of Mars while the scientific instruments on board the orbiter studied the Martian surface, atmosphere and interior. The orbiter made numerous findings during the nine years in which it was operational. *MGS* confirmed that Mars has no significant global magnetic field, supporting the earlier findings of *Mariner 4*. Instead, it discovered that the magnetic field is localised in small areas of the crust. Daily wide-angle images of Mars provided a record of the changing weather conditions on the planet and the orbiter observed that dust storms repeat in the same location year after year. From its lofty vantage point, it spotted changes taking place on the surface, including the formation of new impact craters.

NASA's *Mars Odyssey* and the European Space Agency's (ESA) *Mars Express* spacecrafts arrived at Mars in 2001 and 2003 respectively. *Mars Odyssey* is designed to investigate the Red Planet's environment, mapping its chemical and mineralogical composition and to gather key information about the radiation hazards future crewed missions will face. Early on in its mission, *Mars Odyssey* detected large amounts of hydrogen in the Martian soil, which suggested that water ice lay around a metre below the ground. The spacecraft also found surface deposits of salt in hundreds of locations, indicating that water was once present in those areas and making them targets in the search for evidence of past life. Speaking of the search for life, while the surface of Mars may be inhospitable today, some dark circular features identified by the spacecraft could mark places where life might survive. They are thought to be entrances to caves, and caves would be ideal for sheltering from harsh conditions. Subsurface dwellings would provide protection from solar and cosmic radiation, as well as dust storms, and they would also most likely be able to maintain microclimates that could support life. If these features do turn out to be caves, they may also serve as future shelters for astronauts sent to Mars. *Mars Odyssey* is still going strong and holds the record for the longest continually active spacecraft in orbit around another planet in our Solar System.

Mars Express consisted of two parts: an orbiter and a lander, called *Beagle 2*. On 19 December 2003, just over six months after launch and with the spacecraft nearing the planet, *Beagle 2* was released from the orbiter. On Christmas Day 2003, the orbiter successfully entered orbit around Mars and the lander entered the Martian atmosphere, beginning its descent to an equatorial region called Isidis Planitia, one of the largest impact basins on Mars' surface. *Beagle 2* was to be the first mission since the *Viking* landers to search for life on Mars, but no signal was received from the lander to indicate that it had touched down safely. Mission scientists were unable to determine what had happened at the time, but the answer would eventually come (as you'll discover later on).

The purpose of the *Mars Express* orbiter is to study many aspects of Mars, including its atmosphere, geology and the mineral composition of the surface. Since entering orbit, it has made a number of discoveries including detecting methane in the atmosphere, the first direct observations of carbon-dioxide ice clouds and the first direct detection of **hydrated minerals** (named for the fact that they contain water in their crystalline structure) such as clay. The latter suggests that large, stable amounts of liquid water once existed on the surface of the planet for hundreds of millions of years. In addition, *Mars Express* has not only confirmed the presence of water ice at the planet's polar ice caps, but its radar observations have also detected salty liquid water lying beneath the southern ice cap. This is an exciting discovery because these pockets of liquid water could potentially support life. And, if that wasn't enough, the orbiter made the first detection of aurorae on Mars. The aurorae were spotted during night-time observations of the planet's southern hemisphere. Aurorae are a familiar sight not just on our home planet but on other worlds in our Solar System too, and are produced when charged particles from the **solar wind** flow along a planet's magnetic field lines and collide with gases present in its atmosphere. Although Mars does not have a global magnetic field, the observed aurorae are produced through the interaction of the solar wind and the crustal field regions.

The *Pathfinder* mission ushered in a new era of exploration of the Red Planet. In 2004, *Spirit* and *Opportunity*, twin craft from the NASA *Mars Exploration Rover* programme, touched down on the planet. Each about the size of a golf cart and equipped with an array of instrumentation, these robotic geologists were tasked with finding evidence of past water activity on Mars. The Gusev Crater, an impact basin located in the southern hemisphere, was selected as the landing spot for *Spirit* because its rim had been breached by Ma'adim Vallis, a channel that in Mars' wetter past once carried water into the crater (the breach is estimated to have occurred somewhere between 3.5 and 3.8 billion years ago). *Opportunity* was sent to Meridiani Planum, a plain lying midway between the Tharsis Region and the Hellas Planitia impact basin, near the prime meridian of Mars.[14] The site was chosen because data obtained by the orbiting satellites showed the presence of a mineral called grey crystalline hematite. On Earth, deposits

---

[14] *Opportunity* was halfway around the planet from *Spirit*.

of hematite can form when liquid water circulates through iron-rich rocks.

Planned to last for 90 Martian days (**sols**), the solar-powered twin rovers were operational well beyond this initial mission lifespan and the results obtained far exceeded the expectations of mission scientists, not just in terms of the longevity of the rovers but also the discoveries they made. Returning over 342,000 images of Mars, *Spirit* and *Opportunity* showed us the true beauty of our planetary neighbour. From stunning sand dunes sculpted by the Martian winds, to whirling dust devils[15] and rock-strewn plains, the two rovers transported us to the planet's surface and took us along for the journey. The images were spectacular and so too

---

[15] The rovers sometimes found themselves in the paths of dust devils. No damage was caused though – the dust devils were actually beneficial and cleaned the dust that had accumulated on their solar panels!

were the scientific results. Both rovers found compelling evidence that Mars was once a much wetter planet with conditions that could have supported microbial life, had it existed.

Opportunity discovered small spheres of hematite, nicknamed 'blueberries', that were formed from rising groundwater.[16] Minerals in the water slowly precipitated out over time, producing the spheres we see today. At Endeavour Crater, the rover spotted bright veins of gypsum, a mineral associated with the evaporation of sea water on Earth, that were most likely deposited by water. It also detected clay minerals, which, as we have found on our home planet, tend to form in

---

[16] Scientist Steve Squyres, who was the head of the *Opportunity* science team, said the spheres looked like 'blueberries in a muffin' and so the name stuck. In false-colour images the spheres appear blue too, so the nickname is appropriate!

pH-neutral water (water that is neither acidic nor alkaline). Not to be outdone, *Spirit* revealed that an **outcrop** called Comanche was rich in magnesium and iron carbonates. On Earth, carbon dioxide is trapped in carbonate rocks, so the discovery of carbonates on Mars indicates that the planet's atmosphere was once much richer in carbon dioxide. These rocks formed when Mars was a warmer and wetter planet. *Spirit* also found evidence of ancient volcanism, as well as pure silica, which usually exists in hot springs on Earth. Active volcanism and hot springs may conjure up images of conditions too extreme for life to exist, but, as we know from our home world, life can survive even in the harshest of conditions.

Although continuing for longer than intended, all good missions must come to an end. In May 2009, *Spirit* became stuck in soft soil and, despite the best efforts of engineers to help the rover free itself, it was unable to move. As an added blow, the rover could not align its solar panels to receive enough sunlight to see it through the coming Martian winter. Communication with *Spirit* ceased in March 2010 and in 2011 NASA officially ended its efforts to re-establish contact. *Opportunity* continued exploring the Red Planet until 2018. In the beginning of June that year, localised dust storms merged and began to spread, enshrouding the planet in a global dust storm. Such storms on Mars are not unusual and occur roughly once every six Earth years, but they can play havoc with solar-powered rovers and landers. The storm prevented sunlight from reaching *Opportunity*'s solar panels and meant that the rover was unable to charge its batteries. Mission scientists received the last signal from *Opportunity* on 10 June 2018 and although the global dust storm had cleared by mid-September that year, no further communication was received from the rover. It does, however, currently hold the record for the longest distance travelled by a vehicle on a world other than our own, a total of 45.16 km.

In 2006 the *Mars Reconnaissance Orbiter* (*MRO*) joined the fleet of orbiting spacecraft. Equipped with a suite of instrumentation including a camera called HiRISE (High-Resolution Imaging Science Experiment), *MRO* was tasked with producing a high-resolution map of Mars. The HiRISE camera consists of a reflecting telescope 0.5 m in diameter – the largest telescope ever sent to another planet. To give you an idea of just how good the resolution of the HiRISE camera is, it is capable of seeing features as small as the size of a kitchen table on the surface of the planet – that's pretty impressive considering the altitude of its orbit ranges from 250 km to 316 km above the planet's surface. This resolution means it can easily keep an eye on the robotic explorers on the ground. As a result, in 2015, scientists were finally able to confirm that the mystery of what happened to *Beagle 2* had been solved. The lander had been identified in images taken by *MRO* in 2013 and 2014 and subsequent analysis showed that, although *Beagle 2* had survived the landing, some of its solar panels failed to deploy, which prevented it from phoning home.

HiRISE has returned millions of photographs of Mars to scientists on Earth and has captured dramatic scenes from orbit including active avalanches, soaring dust devils and fresh impact craters (see image 7). Scientists have used the images to estimate that the planet is bombarded by more than 200 small asteroid or comet pieces each year, producing craters at least a few metres in diameter. Its observations of some of the oldest and most heavily cratered regions on the planet have shown that diverse types of wet environments existed in Mars' past and that some would have been more favourable for life than others. With the results of *MRO*'s investigations, scientists are hard at work determining landing sites for future missions and both *MRO* and *Mars Odyssey* currently act as communication relay stations for the robotic explorers on the Martian surface.

Although great strides in the exploration of Mars had been made by the early 2000s, there remained regions untouched by and unfamiliar to robotic explorers. When NASA's *Phoenix Mars* lander touched down near the Martian north pole in 2008 it became the first spacecraft to explore this freezing region. Sampling the soil at its landing site, the lander verified the presence of water ice in the subsurface of Mars and discovered salts that could be nutrients for life.

Something the lander spotted that hadn't been seen on Mars before was carbon-dioxide snow falling from Martian clouds. Unfortunately, but not unexpectedly, contact with the *Phoenix* lander was lost in November 2008 as winter arrived at the landing site. The realities of the harsh conditions at the Martian poles came to light in 2010 when HiRISE images of the lander revealed that one of the delicate solar panels had broken off, most likely due to the build-up of carbon-dioxide ice on its surface.

Of all the missions sent to Mars, just over half have been successful. While it's a success rate that might deter some, the lure of Mars proves too hard to resist and missions to the planet have become increasingly ambitious.

Seven Minutes of Terror

NASA's *Curiosity* rover, part of the *Mars Science Laboratory* mission, was a huge technological leap forward in our studies of Mars when it launched in 2011. Around the size of a car and equipped with an array of advanced scientific instrumentation, the rover's weight was just a tad greater than the weight of the *Mars Exploration* rovers, coming in at 900 kg compared to the 185-kg weights of *Spirit* and *Opportunity*. The sheer size and mass of *Curiosity* meant that the airbag landing system used for the previous rovers was not practicable and an entirely new landing system had to be developed to get the rover safely onto the surface of the planet.

If the idea of using airbags to land a rover on Mars made you break out into a cold sweat, you might need to lie down after reading about the system invented for *Curiosity*. Scientists had settled on the idea of using a rocket-powered 'sky crane', which, after parachutes had slowed the descent of the rover sufficiently, would gently lower it onto the surface of the planet using cables (see Figure 10). Once

Figure 10. The new sky-crane system allows for the delivery of much heavier payloads to Mars. *NASA/ JPL-Caltech*

safely on the ground, the cables would be cut and the sky crane would fly off and crash-land a safe distance from the rover. If the sky crane failed, the *very* expensive rover would smash into the Martian surface. No pressure at all…

'Seven minutes of terror' is the description used by mission scientists to refer to the entry, descent and landing phases of a Mars mission.[17] On 5 August 2012 nerves were at an all-time high as mission scientists waited for confirmation that *Curiosity* had landed. *MRO* had been carefully placed into an orbital position that would allow it to capture *Curiosity* as it descended to the ground. If no landing confirmation was received, the scientists could use the *MRO* data to figure out what had gone wrong. With *Mars Odyssey* acting as a relay station,

---

[17] Hopefully that description doesn't end up matching the time taken to read this chapter and how you feel while reading it!

1. The first colour image of Mars, manually assembled by the *Mariner 4* mission scientists. Pastels were used to hand-colour the image. *NASA/JPL-Caltech/Dan Goods*

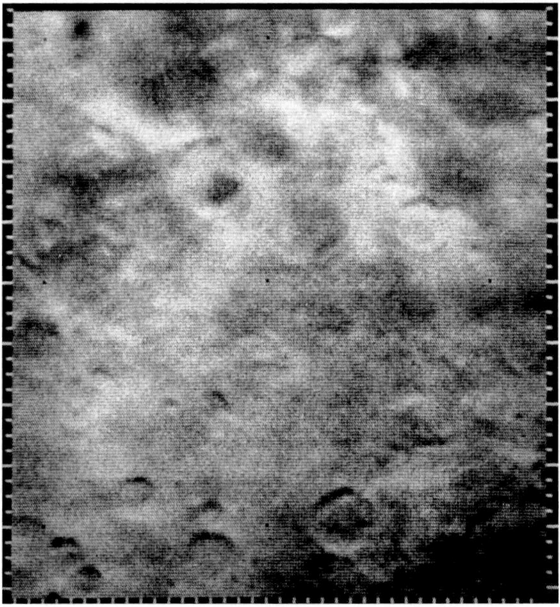

2. A *Mariner 4* image showing craters on the surface of the planet. *NASA/JPL-Caltech*

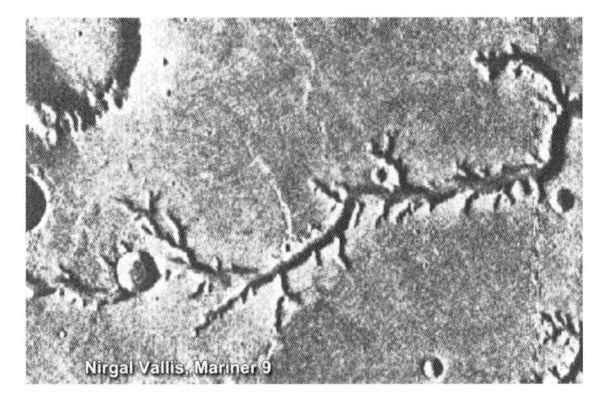

3. The Nirgal Vallis river valley on Mars as imaged by *Mariner 9*. *NASA/JPL-Caltech*

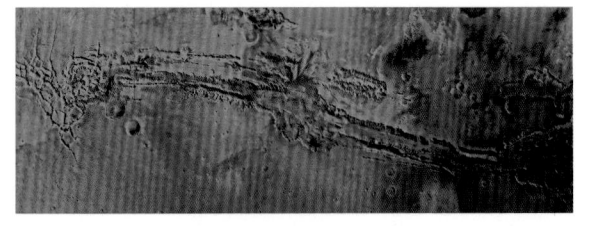

4. A *Viking 1* view of Valles Marineris, the largest canyon system in the entire Solar System. *NASA/JPL/USGS*

5. A rock named 'Big Joe' was captured in this view from the *Viking 1* landing site. *NASA/JPL*

6. The *Sojourner* rover next to a rock called 'Yogi'. Note the deflated airbags surrounding the *Pathfinder* lander. NASA/JPL.

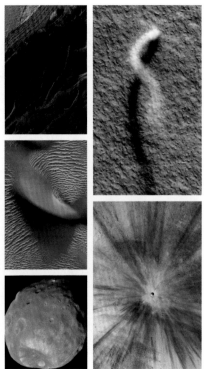

7. The *Mars Reconnaissance Orbiter* has provided stunning views of the Red Planet and its moons. Clockwise from top left: A swirling dust devil gliding along the surface; the dramatic sight of a new crater formed by a meteor impact, with subsurface material scattered across the ground; Phobos, the largest natural Martian satellite; a false-colour image of a large sand dune and wind-sculpted sand ripples; an avalanche captured in action as cliffs near Mars' north pole began to crumble due to seasonal changes. *NASA/JPL-Caltech/University of Arizona*

8. *Curiosity* snapped this view showing evidence of an ancient lake in Gale Crater. The colour has been adjusted so that the rocks look like they would on Earth. *NASA/JPL-Caltech/MSSS*

9. A selfie of *Perseverance* and *Ingenuity* on the surface of Mars. NASA/JPL-Caltech/MSSS

10. A *Mars Global Surveyor* image showing water-ice clouds floating above the Tharsis volcanoes. The giant canyon system Valles Marineris can be seen too, as well as the northern polar ice cap. *NASA/JPL/MSSS*

the descent data was carefully scrutinised as it came through for confirmation that the sky crane had indeed worked. With the words 'Touchdown confirmed! We're safe on Mars!', elation swept through mission control. The next-generation rover *Curiosity* was on the Red Planet.[18]

Just as the soil and rocks on Earth contain the climate and geological history of our home planet, so too do those on Mars. By taking samples and determining their chemical composition and formation processes, we gain insights into the planet's past. *Curiosity*'s design allows it to scoop up soil and small rocks for analysis and it has been equipped with a drill so that it can obtain samples from larger rocks too. For rocks that are out of reach or terrain too treacherous

---

[18] Due to the number of time zones in North America, the landing date is given as both 5 August 2012 (PDT) and 6 August 2012 (EDT).

to drive towards, a laser can be targeted to vaporise tiny amounts of material and reveal its chemical composition. One of *Curiosity*'s objectives is to search for the chemical building blocks of life, namely carbon, hydrogen, nitrogen, oxygen, phosphorus and sulphur. Just like *Spirit* and *Opportunity*, *Curiosity* is equipped with a camera to capture each step of its journey on Mars (see image 8).

Curiosity set down in Gale Crater and set about exploring its new home. It wasn't long before it made its first big discovery – evidence of an ancient river whose water may have been hip-deep in places. On the riverbed were smooth and rounded pebbles that had most likely rolled downstream. When sampling powder drilled from mudstone in a region called Yellowknife Bay, located around 0.5 km from its landing site, *Curiosity* found the ingredients for life we had been hoping it would uncover all along, providing more evidence that conditions on ancient Mars could once have supported microbes. Perhaps most excitingly, *Curiosity* detected a seasonal variation in methane concentrations in the Martian atmosphere. In looking for signs of life and life-supporting environments, methane is a gas that scientists search for because it can be produced by living organisms as a waste product. Further investigation is, however, needed as methane is also a product of chemical reactions between rock and water, and we don't yet know which process is responsible for the methane on Mars.

During its time on the planet, *Curiosity* has explored many fascinating regions and has travelled over 25 km. It is still in operation at the time of publication and busy exploring Mount Sharp, a mountain of sediment in Gale Crater that rises 5.5 km above the crater floor. Mount Sharp's rock layers (**strata**) preserve a geological record of Mars that goes back billions of years. By taking and analysing samples of the layers present in Mount Sharp from its base upwards, *Curiosity* will be able to create a picture of how Mars' environment has changed with time. Will *Curiosity* make it all the way to the top? We'll have to wait to find out!

The number of operational satellites in orbit around Mars has been steadily increasing since scientists sent the first spacecraft in its direction in the 1960s. In 2014, NASA's *Mars Atmosphere and Volatile Evolution* (*MAVEN*) orbiter began its mission to study the structure and composition of the planet's upper atmosphere and investigate how the Martian atmosphere interacts with the Sun and the solar wind. Within weeks of arriving at Mars, *MAVEN* observed that the solar wind was actively stripping the planet of its atmosphere through a physical process called 'sputtering'.

Ions (electrically charged particles) are carried along by the solar wind and slam into the top of the Martian atmosphere, kicking carbon dioxide molecules as well as atoms such as oxygen, carbon and hydrogen into space. This discovery has led to a greater understanding of how the climate of Mars changed over time. How? I'll reveal all in a little while.

Another orbiter interested in the atmosphere of Mars is the ESA *ExoMars Trace Gas Orbiter (TGO)*, which reached the planet in 2016. Water is found in the Martian polar ice caps and buried underground but can escape from these regions, rising into the atmosphere in the form of vapour before being lost to space. *TGO* will track the water-vapour loss and how it is linked to seasonal changes. The orbiter will also search for methane and other gases in the Martian atmosphere as evidence of possible geological or biological activity on the planet. Surprisingly, despite detections by other spacecraft previously, *TGO* has yet to identify methane in the Martian atmosphere.[19] It has, however, identified a new gas not seen on the planet before: hydrogen chloride. The gas was spotted during the global dust storm in 2018[20] and is believed to have formed in the atmosphere as a result of interactions between salt in the fine Martian dust and water.

While most missions to Mars have focussed on exploring the surface and atmosphere of the planet, one mission was sent to explore its interior. NASA's *InSight* lander touched down on Mars in 2018 and was the first spacecraft to robotically place a seismometer on the surface of another planet. After calibrating the instrument, *InSight* began listening

---

[19] This is the case at the time of publication in autumn 2022, anyway!

[20] RIP *Opportunity*.

for seismic vibrations within Mars. These 'marsquakes' have been used to probe the planet's interior. As seismic waves pass through the interior of Mars, the speed and shape of the waves change depending on the properties of the material they pass through. Using the recorded seismic data, scientists have been able to determine that the Martian core has a diameter of just under 3,700 km, about half of the size of the planet itself. The data returned by the lander has also confirmed that the centre of Mars is molten. *InSight* will continue to listen for marsquakes but the fine Martian dust is already beginning to take its toll on the lander, accumulating on its solar panels. If it's lucky, a passing dust devil might help prolong its life on the Red Planet.

February 2021 was a busy month for new Mars missions. The first to reach the planet was the *Hope* orbiter, the United Arab Emirates' first interplanetary spacecraft, designed to build a complete picture of the Martian atmosphere and its layers. The orbiter will study how the Martian atmosphere changes on a daily and seasonal basis. So far, the *Hope* orbiter has observed Martian dust storms billow across the planet's surface and spotted the daily cycles of **clouds of water ice** around the planet, watching how they grow at the beginning and end but shrink during the middle of the day. Returning spectacular global views of Mars, it has also taken the most detailed images so far of its aurorae.

The second mission, and China's first to the planet, that arrived during this period was *Tianwen-1* (meaning 'questions to heaven'). Consisting of an orbiter accompanied by a rover called *Zhurong* ('God of Fire'), the *Tianwen-1* mission will explore the geology of Mars and search for water and ice. The orbiter will use a radar to probe structures more than 100 m below the polar ice caps to learn more about the subsurface liquid water first identified by *Mars Express*. Meanwhile, *Zhurong* made it safely to the surface of Mars in May 2021 and China became only the second country in history to put a rover on the planet. *Zhurong* will survey Utopia Planitia, studying the region's surface and subsurface, as well as searching for evidence of **permafrost** beneath the ground. By May 2022, it had already travelled over 1.9 km and found evidence that its landing site has experienced long periods of weathering due to both water and wind. Scientists believe that Utopia Planitia was host to a massive ocean billions of years ago and the results obtained by *Zhurong* may confirm whether or not this was the case.

NASA's *Mars 2020 Perseverance Rover* was the third mission to arrive at the planet in 2021. The rover wasn't alone on its journey and carried an 'ingenious' payload with it to Mars.[21] *Perseverance* is the most advanced NASA rover ever sent to Mars and the heaviest too, weighing in at 1,025 kg (see Figure 11). Employing the same sky-crane method used to deliver *Curiosity* to the Red Planet's surface, *Perseverance* touched down safely in Jezero Crater to begin its search for signs of ancient life.[22] Jezero Crater is an ideal location for this as satellite observations showed that it is home to an ancient river delta formed from the sediment deposited by flowing water. Similar to *Curiosity*, *Perseverance* has a drill that allows it to collect samples of Martian rock and soil, but what sets it apart is that it will store rock samples for collection in the future by a sample return mission.

---

[21] See what I did there? Need another clue?

[22] Cameras onboard *Perseverance* captured dramatic footage of the landing sequence, which was released to the public. If you haven't seen it, you really should.

Figure 11. The evolution of NASA Mars rovers. This artwork is etched onto a metal plate on the *Perseverance* rover. *NASA/JPL-Caltech*

By summer 2022 *Perseverance* had travelled more than 10 km in the vicinity of its landing site and had already made a number of findings (see image 9). Analysis of a rock sample suggested that parts of the crater floor were likely formed from magma deposited by an ancient lava flow, while studies of rocks found in the crater showed that they had undergone multiple interactions with water over long periods of time. Using SHERLOC (Scanning Habitable Environments with Raman & Luminescence for Organics & Chemicals), one of its onboard instruments, *Perseverance* has discovered organic compounds in the interiors of rocks.[23] Finding organic compounds is by no means confirmation that life once existed because non-biological mechanisms can produce these compounds as well. What is exciting, though, is that if organic compounds have been preserved inside these rocks, signs of life, whether past or present, could be preserved within them too.

In looking towards future crewed missions to Mars, *Perseverance* has successfully made oxygen using carbon dioxide from the Martian air. In its first test, around 5 g of oxygen was made by an instrument called MOXIE (Mars Oxygen In-Situ Resource Utilization Experiment). While this might not sound like a lot, it would be enough to provide

---

[23] WATSON is on Mars too – it's a camera that supports SHERLOC.

an astronaut with around ten minutes of breathable air. Further tests will be carried out, but scientists think that larger versions of the MOXIE instrument could be taken to Mars by crewed missions. Creating oxygen from the air on Mars would reduce the amount of oxygen that would need to be carried from Earth to the Red Planet to sustain a mission of this type.

Not only are we interested in the sights of Mars, but its sounds can also teach us about the Martian environment. Two microphones on board *Perseverance* have captured elements of the planet's soundscape: Martian winds, the noise of the rover's wheels rolling across the surface, as well as the sounds made when *Perseverance* fires its lasers at rocks.[24] By analysing the recordings, scientists have discovered that Mars has not one

---

[24] I know what you're wondering and the answer is no, the laser does not make 'pew pew' sounds.

but two speeds of sound. On Earth, sound typically travels at 343 m per second (m/s) through the air but on Mars scientists have calculated that low-pitch sounds travel at around 240 m/s while high-pitch sounds travel slightly faster at around 250 m/s. This variation in speed is due to the thin carbon-dioxide atmosphere on Mars and could potentially cause communication problems for astronauts. It would mean our ears would hear the high-pitched sounds slightly earlier than the low-pitched sounds. The sound level on Mars is lower too so you would need to be much closer to the sound source to hear it at the same volume as you would on Earth. If you're having a conversation with a fellow astronaut, get around two car lengths away from them and, without a radio, you won't be able to hear what they are saying.

With every successful mission to Mars, we are slowly piecing together the planet's history. Seeing all the amazing images taken from above and on the ground, it's difficult to picture Mars as a world that once had oceans and water flowing on it. It does make you wonder, though. How did Mars go from being a warm and potentially hospitable planet to the cold and dry world we see today?

Rocky III vs. Rocky IV

Mars was doomed to fail as a planet from the very beginning. As an opening line, this is a rather gloomy statement to make, but recent research suggests that Mars' fate was indeed sealed early on. After the chaotic and violent formation of our Solar System around 4.6 billion years ago, Mars and the other rocky planets were balls of molten rock. Eventually, they started to cool down. Mars' atmosphere formed as a result of **outgassing** of the **mantle**, with asteroid and possibly comet impacts also making contributions. Water vapour in the atmosphere eventually condensed and began to fall to the surface, forming the first oceans.

Space rocks pummelled the surface of the Red Planet, heavily cratering its southern hemisphere and producing the Hellas, Isidis and Argyre impact basins. Volcanism led to the growth of the Tharsis region, home to the volcanoes Arsia Mons, Pavonis Mons, Ascraeus Mons and Olympus Mons.[25] As this region grew, the planet's surface fractured, creating the vast canyon system Valles Marineris, which covers an expanse of 4,000 km and reaches depths of 7 km. The volcanoes pumped gases into the Martian atmosphere making it thicker and denser, which allowed more solar heat to be trapped and warmed the planet in turn. This heating was much needed because Mars orbits the Sun at an average

---

[25] Olympus Mons is the highest mountain and volcano in the Solar System, rising about 25 km above the Martian surface.

distance of 228 million km, placing it on the outskirts of the **habitable zone**. This region, also sometimes referred to as the 'Goldilocks zone', is the region around a star where liquid water could exist on a planet's surface, provided the planet has a suitable atmosphere. In our Solar System, conservative estimates place the inner and outer edges of the habitable zone at around 0.95 to 1.7 au from the Sun respectively, where 1 au (**astronomical unit**) is the average distance between the Sun and Earth (about 150 million km). Earth is comfortably within the habitable zone of our parent star.

So far, so good. Mars was ticking all the boxes needed to become a habitable planet – it was warm and it had liquid water on its surface. But then everything changed. Up until around 4 billion years ago, Mars had a strong global magnetic field that was created by the movement of molten metals in its core. Earth has the same mechanism and this generates our planet's own magnetic field, which, along with the atmosphere, shields us – for the most past – from the bombardment of cosmic and solar radiation. Unluckily for Mars, this protective mechanism switched off. Why it switched off is still a mystery, but what we do know is that the loss of the planet's global magnetic field meant it wasn't long before the solar wind began mercilessly stripping gas from the Martian atmosphere. As a double whammy, recent research suggests that Mars had already been losing its atmosphere prior to the loss of its global magnetic field due to its lower gravity, which makes it easier for gases to escape into space.

Mars is roughly half the size of Earth, with a **mass** about 11% that of Earth.[26] Since mass matters when it comes to gravity,

---

[26] Rocky III (Earth) comes in at 5.972×10^{24} kg while Rocky IV (Mars) comes in at 6.417×10^{23} kg. (Or 5,972,000,000,000,000,000,000, 000 kg versus 641,700,000,000,000,000,000, 000 kg)

Mars' lower mass means its gravitational pull is weaker than Earth's. To give you an idea of just how weak it is, the surface gravity on Mars is only around 38% of the surface gravity on Earth, so if you were to stand on Mars you would feel considerably lighter on your feet. You would also be able to jump around two-and-a-half times higher on Mars than you can here on Earth. This is, of course, great news for any aspiring basketball players or high jumpers who find their true sporting potential thwarted by Earth's gravitational pull, but for Mars it contributed to the catastrophic climate change we touched on earlier.

The loss of a huge portion of the Martian atmosphere led to the planet drying out and cooling, and the lack of a global magnetic field, combined with the thin atmosphere, enveloped the surface in damaging ultraviolet radiation emitted by the Sun. Just how much atmosphere has Mars lost over time? We now have an estimate thanks to the *MAVEN* orbiter. As we've noted, *MAVEN* has observed how the solar wind continues to strip Mars' atmosphere today and discovered that when Mars is at its closest to the Sun the planet loses ten times more hydrogen than it does when it is furthest from the Sun. Analysis of the data obtained by *MAVEN* indicates that Mars has lost about ⅔ of its atmosphere since the planet formed – it turns out that early Mars once had an atmosphere as thick as that of our home planet today. If early Mars was similar to today's Earth, could life have evolved on the planet? If it did, could that life have been driven underground as the surface became increasingly inhospitable?

When we talk about Mars being inhospitable, it's useful to draw comparisons with our home planet to understand just how harsh the conditions are. You can think of the next few paragraphs as the number of ways that

Mars can kill you. Let's begin with the composition of the Martian atmosphere.

Earth's atmosphere consists predominantly of nitrogen (78%) and oxygen (21%) with a mixture of other gases making up the final 1%. Mars' atmosphere is 96% carbon dioxide, with nitrogen, argon and other gases comprising the remaining 4%. Humans and carbon dioxide don't get along all that well, so a spacesuit is strongly recommended if you are going to venture out onto the Martian surface.

We know that the atmosphere on Mars is thin, but how thin exactly? To give you an idea, we can look at the surface air pressure of the planet. On Earth, the air pressure at sea level is 1,013.25 millibars (or about 14.7 pounds per square inch (psi)), whereas on Mars the air pressure on the surface is a paltry 6.5 millibars (or roughly 0.1 psi). At an altitude of 45 km above the surface of

Earth the air pressure is similar to that of Martian surface air pressure. If 45 km from Earth's surface doesn't seem all that high, I encourage you to watch the videos of the frankly terrifying jumps made by skydiver Felix Baumgartner and computer scientist Alan Eustace (both of which started *under* 45 km above Earth). What implication does the low surface pressure on Mars have for humans who want to explore the planet but forget to wear a spacesuit? If you're squeamish, you may want to skip straight to the next paragraph. When atmospheric pressure decreases, so too does the boiling point of a liquid. Pure water, for example, boils at 100 °C at sea level on Earth, but a temperature of about 68 °C is all that is needed for the same to happen at the summit of Mount Everest (8.848 km above sea level). On Mars, however, fresh water starts boiling at 10 °C. The human body contains a lot of water – in terms of a percentage, about 60% of your body is water. Direct exposure to the low atmospheric pressure on Mars will cause the saliva on your tongue and the tears in your eyes to boil away. Gas bubbles will start to form in bodily fluids (the technical term is 'ebullism'). What this all means is that if you did not wear a pressurised spacesuit on the Red Planet, you would fizz to death.[27]

If you are not a cold-weather person, then Mars is not the planet for you. Earth has an average surface temperature of 14 °C, positively tropical compared to the average temperature on Mars, which is a chilly -63 °C. It can get quite cold on our planet, though – the lowest temperature on record is -97.8 °C! As expected, Mars does get a bit chillier than that, with temperatures dropping as low as -140 °C. In terms of summer weather, temperatures on Mars can reach

---

[27] Don't worry – your head will not explode.

as high as 30 °C – rather lovely. If you are curious about the planet's weather, have a look at the daily reports returned from the *Perseverance* and *Curiosity* rovers. In May 2022, Martian mid-autumn for *Perseverance*, its weather report showed a minimum temperature of -80 °C and maximum of -15 °C at Jezero Crater. Over at Gale Crater, it was mid-spring during the same period and *Curiosity* reported a low of -67 °C and a high of 4 °C. The freezing temperatures might make you consider standing in the sunshine on the Martian surface to warm up a bit, but just remember what we covered earlier – you'll be subject to dangerously high levels of radiation. So not only will you need a spacesuit to prevent the fluid in your lungs from vaporising, but it's also highly recommended to keep you warm and protect you from the harmful UV rays.

Seeing a planet that was once so similar to our own as a cold and dry world today shows us just how lucky we are and serves as a stark reminder of the fragility of our own home (see image 10). The discovery of liquid water beneath the Martian surface is promising, though, in our search for life on the planet. *Curiosity*, *Zhurong* and *Perseverance* will continue to explore, looking for signs of past or present life and further missions will join them in the future. The ESA *Rosalind Franklin* rover, part of the *ExoMars* programme, will drill to a depth of 2 m to collect samples to analyse and seek out evidence of life buried underground. And those rock samples that *Perseverance* has been collecting and storing? The *Mars Sample Return* mission is a joint venture between NASA and ESA to collect the samples and return them to Earth. Scientists are eyeing a launch in the mid- to late 2020s with sample return expected in the early 2030s. The combined efforts of the rovers could finally provide the answers to our questions.

It's not just the planet itself that will be the focus of future missions, its moons

Phobos and Deimos will be explored too. Both moons orbit Mars considerably closer than the Moon orbits Earth. The Moon is, on average, 384,000 km from Earth whereas Deimos orbits Mars at an average distance of just over 23,450 km. As for Phobos, it's even closer to the Red Planet and orbits at an average distance of 9,376 km. No other moon in the Solar System orbits its planet as close as Phobos orbits Mars, and it is getting closer and closer to the planet. Approaching Mars at a rate of around 1.8 m per century, Phobos is heading for destruction, albeit a long, long time from now (in 50 million years' time, in case you're wondering). It will either crash into Mars, or the Red Planet's gravity will rip it apart and the debris will go on to form a Martian ring system. The origins of the potato-shaped moons remain unknown. One line of thought is that Phobos and Deimos are captured asteroids and were originally members of the asteroid belt, nudged out of their original home by Jupiter's immense gravity. A different hypothesis is that the two moons we see today are the remains of a shattered moon. The *Martian Moon eXploration (MMX)* mission, designed by the Japan Aerospace Exploration Agency (JAXA), will focus on determining their origin and will also collect a sample to return to Earth. Mission scientists think that Phobos may contain Martian material thrown out during large asteroid and meteorite impacts, so could Phobos have evidence of past life on Mars?

Perhaps the greatest future mission to Mars, though, is one that people have dreamed about, written stories about and featured in films for decades. It's one that has been a long time in the making and will be as historic and profound as *Apollo 11* was for our exploration of the Moon: the first crewed mission to Mars.

Dare Mighty Things

Encoded into the parachute used to slow down the *Perseverance* rover as it made its way to the surface of Mars were the words 'Dare Mighty Things', the motto of NASA's Jet Propulsion Laboratory (JPL).[28] No saying is perhaps truer of the human spirit and our drive to push the boundaries of what is possible. History is littered with examples of humans daring mighty things and conquering the obstacles that were once considered to be beyond our capabilities.

---

[28] Binary code was used to encode two messages into the parachute. The first was the motto and the second message contained the GPS coordinates of JPL.

On 17 December 1903, at Kill Devil Hills near Kitty Hawk, North Carolina, USA, brothers Orville and Wilbur Wright made history when they achieved the first powered flight of an aircraft. Their heavier-than-air airplane, called the Wright Flyer, lifted off its launching rail and remained airborne for 12 seconds before coming to a rest on the ground roughly 37 m (120 feet) from its launch point. This short flight changed the world forever. In a demonstration of just how far technology had progressed since that momentous day, in 1969 humans set foot on the Moon for the first time. In honour of the Wright brothers' achievements, the *Apollo 11* crew carried some pieces of the Wright Flyer with them to the Moon. And now the legacy of Orville and Wilbur Wright extends all the way to Mars.

NASA's *Mars Helicopter*, *Ingenuity*, which travelled to Mars along with the *Perseverance* rover, became the first aircraft to make a controlled, powered flight on another world in our Solar System on 19 April 2021. Climbing to a maximum altitude of 3 m (10 feet) and maintaining a stable hover for 30 seconds, the solar-powered *Ingenuity* achieved a goal once thought impossible. The site where *Ingenuity* performed its first flight has since been named 'Wright Brothers Field' and, as an added tribute, a 1-inch square fragment of the Wright Flyer is now on Mars, safely stowed on board *Ingenuity*.

The incredibly thin Martian atmosphere makes it especially challenging to achieve lift, so *Ingenuity* was designed to be light (it has a mass of 1.8 kg). It was also fitted with four custom-made carbon-fibre blades arranged into two rotors that spin in opposite directions at an astonishing 2,500 revolutions per minute (rpm). For comparison, the blades of the average helicopter on Earth rotate somewhere between 400 and 500 rpm. As with the rovers on the ground, *Ingenuity* cannot be controlled in real time and flies autonomously based on commands that are planned and programmed in advance.

Initially intended as a technology demonstrator, *Ingenuity* was still flying high in June 2022. In the period since its launch, it has completed more than 20 flights on Mars and accumulated more than 50 minutes of flight time. The extraordinary achievements of this unique spacecraft are testament to its robust design and the dedication and hard work of the scientists and engineers involved in the mission. *Ingenuity* has been so successful that after completing its fifth test flight in early May 2021 it officially embarked on a new phase to explore how aerial craft could benefit further exploration of Mars. They could be used to perform surveys and scout out potential routes for rovers, working alongside them to identify areas of interest. Looking towards the future, aerial craft and rovers will play a vital role in helping astronauts explore the Red Planet.

Many believe that the first crewed mission to Mars will take place in the early 2030s, but, considering that the last time humans were on the Moon was back in 1972[29], it is understandable that some people are looking beyond that decade as a more realistic target. Part of the challenge of sending astronauts to Mars is making sure we have the technology to get them to the planet, keep them safe on the ground and then return them safely to Earth. For many, a return to the Moon is vital in establishing and testing the technologies needed for this first crewed mission.

Prior to sending astronauts to Mars, we need to ensure we understand what will happen to the human body, and mind, on long journeys to the planet. With

---

[29] As of autumn 2022.

current spacecraft propulsion systems, a 'short' round trip to Mars would take about 21 months – nine to get there, followed by a roughly three-month period on the Red Planet so that Earth and Mars are in the right alignment for the trip back home to be as short as possible – another nine months. That's a long time to be away from loved ones. Astronauts living and working on the International Space Station for extended periods of time can provide some insight into the challenges that astronauts on a Mars mission would face. Simulated Mars missions on Earth will support research into the development of methods and technologies that will prevent and resolve problems future crewed missions to Mars are expected to encounter. These include the psychological effects associated with isolation, being confined to a small space with fellow astronauts and adapting to living on a world so alien to our own. And, although astronauts would be able to keep in touch with loved ones, the communication delays between Mars and Earth will be one more obstacle to overcome.

The spacecraft that will transport the astronauts to Mars must be large enough to carry the equipment and supplies required for the duration of the entire mission, so not only for their time on the planet, but also for their journey there and back too. Unfortunately, the more mass you load into your rocket, the more fuel you will need and the more expensive your mission becomes. One possible solution would be to send supply ships to Mars ahead of time. These could carry the infrastructure – solar panels and habitats, for example – along with some food and medical supplies. Due to the long journey time, the spacecraft will need to be shielded to protect astronauts from the harmful solar and cosmic radiation that permeates the Solar System. As for the astronauts themselves, onboard exercise equipment will be required to maintain their muscle strength and bone density. In a microgravity environment, bones and muscles are no longer needed to support the body's mass so, without appropriate exercise, astronauts would experience muscle and bone loss. The last thing we want is for astronauts to break their bones within minutes of landing on Mars.

Getting to Mars is one thing, getting astronauts safely onto the ground is another. We've seen airbags and sky cranes used for rovers, but delivering a much larger payload to the planet's surface will necessitate an entirely different landing system. Space agencies are currently investigating different landing methods that could be used for future crewed missions. When it comes to a suitable landing site, the ideal location would be one of high scientific value so that the astronauts can continue the search for signs of Martian life – assuming none have been found by this point – and they'll also be tasked with investigating other mysteries of the Red Planet. The first crewed mission would have a number of scientific objectives, one of which would be to demonstrate the feasibility of future longer-term missions to the planet, so the selected location should be one that would offer the crew the best chances of achieving most, if not all, of the mission aims. Preferably, it would also contain enough resources to sustain a subsequent crewed mission. As *Perseverance* has already demonstrated, it would be possible for astronauts to generate their own oxygen on Mars using the Martian atmosphere. Tapping into underground water would be crucial for those astronauts too, both in the search for microbial life and for its potential use for agriculture and in the production of rocket propellant for a return to Earth.[30]

---

[30] Yes, you would be able to grow potatoes on Mars like Mark Watney in *The Martian*. Just don't get stranded…

Living and working on Mars will take some getting used to. Astronauts could live in habitats on the surface or stay inside the rocket in which they arrived. If the former is the preferred option, the environs chosen would need to protect the astronauts from the harsh surface conditions and be self-sustaining, capable of supporting life for lengthy periods without assistance from Earth (see Figure 12). In the distant future, astronauts could live in habitats in underground caves.

While the length of a Martian day is only slightly longer than a day on Earth,[31] the length of a Martian year is substantially longer than a year on Earth (687 days versus 365.25 days). A longer year means the seasons are longer too. Summer in the southern hemisphere of Mars is much warmer than summer in the planet's northern hemisphere. While this

---

[31] The length of a Monday on Mars is yet to be confirmed.

Figure 12. An artist's impression showing astronauts and habitats on Mars. *NASA*

is great news if you want to keep warm, the downside is that the milder weather in the southern-hemisphere summer creates stronger winds, which stir up the biggest dust storms. Unlike the dramatic scenes shown in some movies, the strongest winds on Mars will, in fact, feel like gentle breezes due to the thinness of the Martian atmosphere. While some dust storms last for a few days and only affect small regions of the planet, others can consume the globe and last for months, as was the case with the dust storm in 2018. This is especially problematic if astronauts are to be reliant on solar panels to generate electricity, as a dust storm can cause a significant drop in the amount of sunlight reaching the planet's surface. Lunar dust played havoc with the *Apollo* astronauts, resulting in eye and respiratory irritations. The fine Martian dust could potentially lead to the same problems and, as an added bonus, the dust might be toxic too.

Once the astronauts have completed their mission, there's the small matter of getting them back to Earth. Lifting off from Mars has never been done before. The *Mars Sample Return* mission will be the first to attempt it, but that will be small fry compared to the prospect of launching a much larger payload back into space – another reminder of the importance of developing and testing the technology needed to get to and from Mars before we send humans there.

Considering the multitude of challenges, why would we ever want to send humans to the Red Planet? After all, the unmanned rovers and landers have already done a great job and made extraordinary progress. In addition, they're designed to withstand the conditions on the planet, whereas humans have evolved to cope with conditions on our home planet. There is no denying that rovers and landers have played a vital role in studying Mars, but there is a limit to how much they can achieve. A trained astronaut can conduct field exploration and install (and maintain) complex scientific instrumentation. Based on experience and judgement, they can immediately identify a region of interest to explore whereas a rover would have to wait for instructions from mission scientists on Earth. Rather than looking exclusively at crewed missions to the planet, we would be wise to consider both. A crewed mission working alongside robotic explorers will be instrumental in understanding the complexities of this fascinating world. The success of a first crewed mission may go on to create a sustainable programme for the exploration of Mars. We may even reach a point where humans could establish a permanent colony on the planet. What was once in the realm of science fiction could one day become reality.

As Carl Sagan said, we are wanderers and the next place to wander to is Mars. When the first humans set foot on the surface of the planet, they will embark on humankind's next greatest adventure. It will be our biggest leap into the cosmos and a stepping-stone on the path to many more crewed missions into the Solar System and beyond. The chosen few will continue the pioneering work of those before them in the quest to solve the mysteries of Mars. While their new home will be alien to them, the familiar night sky will provide a connection to those back home. And when the conditions are just right, there will be one very special object they'll get to see, a bright point of light in the Martian night sky – our home planet, Earth. That will be some view.

Glossary

aphelion – the point at which an object moving in an elliptical orbit around the Sun reaches its greatest distance from the Sun.

astronomical unit (au) – a unit of length used by astronomers, which is approximately equal to the average distance between Earth and the Sun (1 au ≈ 150 million km). The exact value of 1 au as defined by the International Astronomical Union is 149,597,870,700 m (149.5978707 million km).

axial tilt – the angle of the rotational axis of a planet measured relative to a line drawn perpendicular to the plane of the planet's orbit around the Sun.

clouds of water ice – clouds composed of particles of water ice (ice formed by water as opposed to another substance).

deferent – in the geocentric model of the Solar System, the deferent was a large circle along which the centre of an epicycle moved.

dust devils – swirling, narrow columns of hot air, typically tens or hundreds of metres tall, that lift dust off the surface of a planet.

eccentricity – a measure, on a scale of 0 to 1, of how much an ellipse deviates from a perfect circle. If a planet has an orbital eccentricity of 0 it is moving in a circular orbit around the Sun. The higher the orbital eccentricity, the more elliptical the orbit.

electron – a subatomic particle with a negative charge.

epicycle – applicable to the geocentric model of the Solar System, the epicycle was a small circle around which a planet was said to move.

geocentric model – a model of the Solar System with Earth at the centre.

habitable zone – the region around a star where the temperature is mild enough for liquid water to exist on the surface of an Earth-like planet.

heliocentric model – the accepted model of the Solar System, where the Sun is at the centre and the planets orbit the Sun

Hohmann Transfer orbit – the most fuel efficient path for a spacecraft to take to travel from one planet to another.

hydrated minerals – minerals that contain water (H_2O) in their crystalline structure. An example of this is gypsum, calcium sulphate dihydrate – its chemical formula is $CaSO_4. 2H_2O$.

ions – electrically charged particles formed when atoms gain or lose electrons.

limb – the outer edge of a celestial body's disc.

mantle – the layer that lies between the core and the crust of a rocky planet.

mass – a measure of the amount of matter in an object. This is different to weight, which is a force and the measurement of the pull of gravity on the object. Mass is measured in kilograms, while weight is measured in Newtons.

micrometeorite – a space rock that is about the size of, or smaller than, a grain of sand.

opposition – the position of a planet when it is exactly on the opposite side of Earth from the Sun.

outcrop – visible exposures of bedrock, the solid rock that lies under a layer of regolith (fragmented and weathered rock debris, soil and sediments), on the surface of a planet.

outgassing – the release of gases that were previously trapped, frozen, absorbed or dissolved in some material. An example of outgassing is the release of gas during a volcanic eruption.

perihelion – the point at which an object moving in an elliptical orbit around the Sun is closest to the Sun.

permafrost – ground that remains frozen for at least two consecutive years.

retrograde motion – the apparent backward motion of a planet relative to the background stars.

sol(s) – unit of timekeeping on Mars. A Martian day is referred to as a 'sol' and is approximately 40 minutes longer than a day on Earth.

solar wind – a stream of fast-moving charged particles emanating from the Sun and flowing outwards through the Solar System.

spectroscopy – the science of obtaining and studying the spectra of objects. By splitting light into its different colours (or wavelengths), astronomers can gather a wealth of information about the object emitting the light, including, for example, its chemical composition.

strata – layers of sedimentary rock (rocks formed from the accumulation of sediments).

technology demonstrator – a mission designed to demonstrate operational techniques and capabilities, and/or technologies and advanced systems that can be incorporated into future missions.

vesicle – a small hole in volcanic rock formed by the expansion of a gas bubble that was trapped inside the lava.